Kakuro Puzzle Book For Adult Relaxation

300 Moderately Easy Puzzles
Volume 1
(Solutions Included)

This Book Belongs To:

Copyright and other intellectual property laws protect these materials. Reproduction or retransmission of the materials in whole or part thereof in any manner, without the prior written consent of the copyright holder, is a violation of the copyright law and will be enforced to the fullest extent of the applicable law.

Copyright © 2019 Backdoor Publishing
All Rights Reserved

What is Kakuro?

Kakuro puzzle, also known as cross sums are numerical crosswords. It is a kind of logic puzzle that is very addictive.

How To Play The Kakuro Puzzle

In this puzzle game, digits are used so as to add these numbers to sum up for values which are stated in the squares of board.
Moreover, within each sum group, digits appear only once.

The conventional method of solving kakuro puzzles is incremental. Using the information seen on the board, you can easily get the value of blank cells which can only have one possible value.

Kakuro instructions are very simple. All you have to do is to arrange the numbers 1 to 9 inside the board grid so that each incessant vertical or horizontal grid of blank squares total into the value on the left of it or above it correspondingly.

I will appreciate positive feedback so as to serve you better in subsequent puzzles.
Thank You.

Solutions from Page 79

Puzzle 1

Puzzle 2

Puzzle 3

Puzzle 4

Puzzle 9

Puzzle 10

Puzzle 11

Puzzle 12

Puzzle 13

Puzzle 14

Puzzle 15

Puzzle 16

Puzzle 17

Puzzle 18

Puzzle 19

Puzzle 20

Puzzle 21

Puzzle 22

Puzzle 23

Puzzle 24

Puzzle 25

Puzzle 26

Puzzle 27

Puzzle 28

Puzzle 29

Puzzle 30

Puzzle 31

Puzzle 32

Puzzle 33

Puzzle 34

Puzzle 35

Puzzle 36

Puzzle 37

Puzzle 38

Puzzle 39

Puzzle 40

Puzzle 41

Puzzle 42

Puzzle 43

Puzzle 44

Puzzle 45

Puzzle 46

Puzzle 47

Puzzle 48

14

Puzzle 49

Puzzle 50

Puzzle 51

Puzzle 52

Puzzle 53

Puzzle 54

Puzzle 55

Puzzle 56

16

Puzzle 57

Puzzle 58

Puzzle 59

Puzzle 60

Puzzle 61

Puzzle 62

Puzzle 63

Puzzle 64

Puzzle 65

Puzzle 66

Puzzle 67

Puzzle 68

Puzzle 69

Puzzle 70

Puzzle 71

Puzzle 72

Puzzle 73

Puzzle 74

Puzzle 75

Puzzle 76

Puzzle 77

Puzzle 78

Puzzle 79

Puzzle 80

Puzzle 81

Puzzle 82

Puzzle 83

Puzzle 84

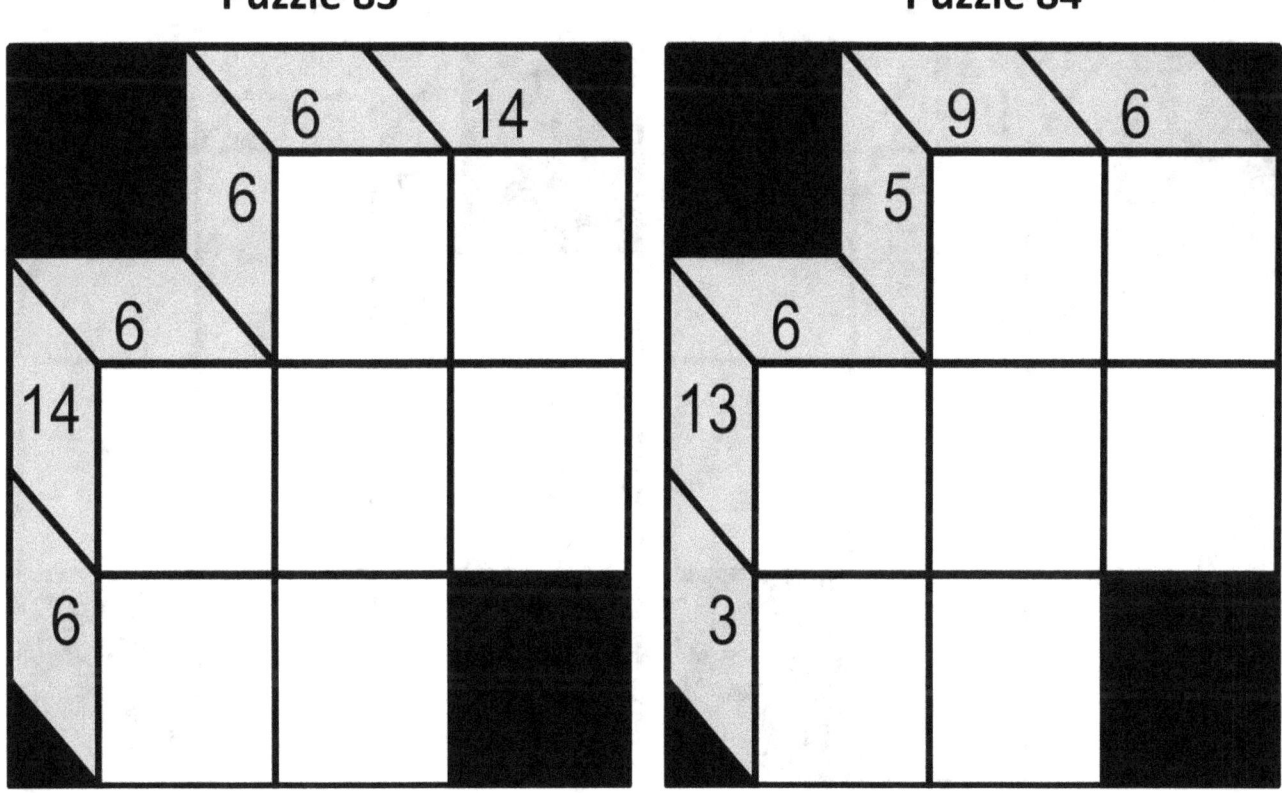

Puzzle 85

Puzzle 86

Puzzle 87

Puzzle 88

Puzzle 89

Puzzle 90

Puzzle 91

Puzzle 92

Puzzle 93

Puzzle 94

Puzzle 95

Puzzle 96

Puzzle 97

Puzzle 98

Puzzle 99

Puzzle 100

Puzzle 101

Puzzle 102

Puzzle 103

Puzzle 104

28

Puzzle 105

Puzzle 106

Puzzle 107

Puzzle 108

Puzzle 109

Puzzle 110

Puzzle 111

Puzzle 112

Puzzle 113

Puzzle 114

Puzzle 115

Puzzle 116

Puzzle 117

Puzzle 118

Puzzle 119

Puzzle 120

Puzzle 121

Puzzle 122

Puzzle 123

Puzzle 124

Puzzle 125

Puzzle 126

Puzzle 127

Puzzle 128

34

Puzzle 129

Puzzle 130

Puzzle 131

Puzzle 132

Puzzle 133

Puzzle 134

Puzzle 135

Puzzle 136

Puzzle 137

Puzzle 138

Puzzle 139

Puzzle 140

Puzzle 141

Puzzle 142

Puzzle 143

Puzzle 144

Puzzle 145

Puzzle 146

Puzzle 147

Puzzle 148

Puzzle 149

Puzzle 150

Puzzle 151

Puzzle 152

40

Puzzle 153

Puzzle 154

Puzzle 155

Puzzle 156

Puzzle 157

Puzzle 158

Puzzle 159

Puzzle 160

Puzzle 161

Puzzle 162

Puzzle 163

Puzzle 164

Puzzle 165

Puzzle 166

Puzzle 167

Puzzle 168

Puzzle 169

Puzzle 170

Puzzle 171

Puzzle 172

Puzzle 173

Puzzle 174

Puzzle 175

Puzzle 176

Puzzle 177

Puzzle 178

Puzzle 179

Puzzle 180

Puzzle 181

Puzzle 182

Puzzle 183

Puzzle 184

48

Puzzle 185

Puzzle 186

Puzzle 187

Puzzle 188

Puzzle 189

Puzzle 190

Puzzle 191

Puzzle 192

50

Puzzle 193

Puzzle 194

Puzzle 195

Puzzle 196

Puzzle 197

Puzzle 198

Puzzle 199

Puzzle 200

52

Puzzle 201

Puzzle 202

Puzzle 203

Puzzle 204

Puzzle 205

Puzzle 206

Puzzle 207

Puzzle 208

Puzzle 209

Puzzle 210

Puzzle 211

Puzzle 212

Puzzle 213

Puzzle 214

Puzzle 215

Puzzle 216

Puzzle 217

Puzzle 218

Puzzle 219

Puzzle 220

57

Puzzle 221

Puzzle 222

Puzzle 223

Puzzle 224

Puzzle 225

Puzzle 226

Puzzle 227

Puzzle 228

Puzzle 229

Puzzle 230

Puzzle 231

Puzzle 232

Puzzle 233

Puzzle 234

Puzzle 235

Puzzle 236

61

Puzzle 237

Puzzle 238

Puzzle 239

Puzzle 240

Puzzle 241

Puzzle 242

Puzzle 243

Puzzle 244

Puzzle 245

Puzzle 246

Puzzle 247

Puzzle 248

64

Puzzle 249

Puzzle 250

Puzzle 251

Puzzle 252

Puzzle 253

Puzzle 254

Puzzle 255

Puzzle 256

66

Puzzle 257

Puzzle 258

Puzzle 259

Puzzle 260

Puzzle 261

Puzzle 262

Puzzle 263

Puzzle 264

Puzzle 265

Puzzle 266

Puzzle 267

Puzzle 268

Puzzle 269 **Puzzle 270**

Puzzle 271 **Puzzle 272**

70

Puzzle 273

Puzzle 274

Puzzle 275

Puzzle 276

Puzzle 277

Puzzle 278

Puzzle 279

Puzzle 280

Puzzle 281

Puzzle 282

Puzzle 283

Puzzle 284

Puzzle 285

Puzzle 286

Puzzle 287

Puzzle 288

74

Puzzle 289

Puzzle 290

Puzzle 291

Puzzle 292

75

Puzzle 293

Puzzle 294

Puzzle 296

Puzzle 297

Puzzle 298

Puzzle 299

Puzzle 300

Solutions Start Next Page

Answer 1

Answer 2

Answer 3

Answer 4

Answer 5

Answer 6

Answer 7

Answer 8

Answer 9

Answer 10

Answer 11

Answer 12

Answer 13

Answer 14

Answer 15

Answer 16

Answer 17

Answer 18

Answer 19

Answer 20

Answer 21

Answer 22

Answer 23

Answer 24

Answer 25

Answer 26

Answer 27

Answer 28

Answer 29

Answer 30

Answer 31

Answer 32

Answer 33

Answer 34

Answer 35

Answer 36

Answer 37

Answer 38

Answer 39

Answer 40

Answer 41

Answer 42

85

Answer 43

Answer 44

Answer 45

Answer 46

Answer 47

Answer 48

Answer 49

Answer 50

Answer 51

Answer 52

Answer 53

Answer 54

Answer 55

Answer 56

Answer 57

Answer 58

Answer 59

Answer 60

Answer 61

Answer 62

Answer 63

Answer 64

Answer 65

Answer 66

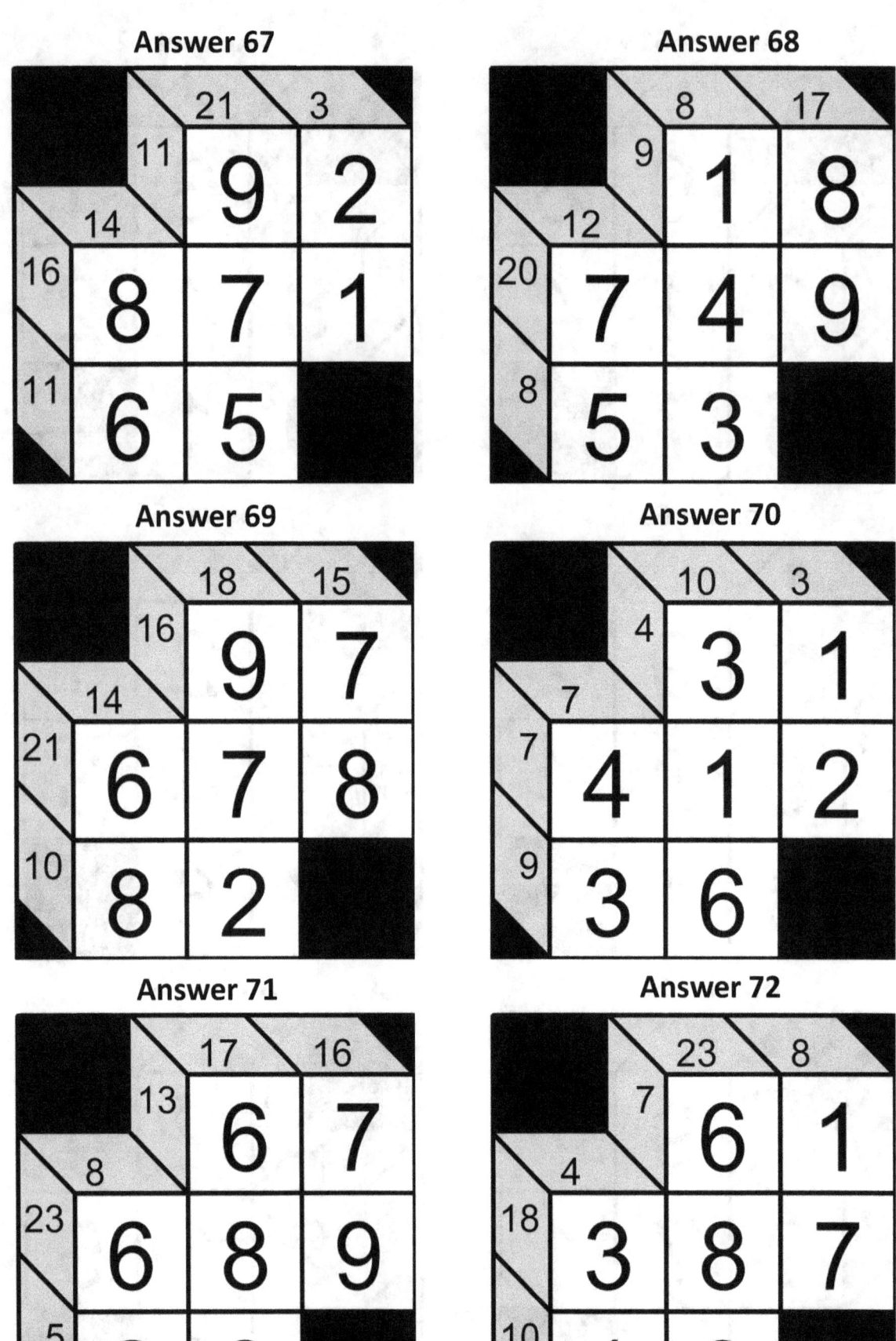

Answer 73

Answer 74

Answer 75

Answer 76

Answer 77

Answer 78

Answer 79

Answer 80

Answer 81

Answer 82

Answer 83

Answer 84

92

Answer 85

Answer 86

Answer 87

Answer 88

Answer 89

Answer 90

Answer 91

Answer 92

Answer 93

Answer 94

Answer 95

Answer 96

Answer 97

Answer 98

Answer 99

Answer 100

Answer 101

Answer 102

Answer 103

Answer 104

Answer 105

Answer 106

Answer 107

Answer 108

Answer 109

Answer 110

Answer 111

Answer 112

Answer 113

Answer 114

Answer 115

Answer 116

Answer 117

Answer 118

Answer 119

Answer 120

Answer 121
Answer 122
Answer 123

Answer 124
Answer 125

Answer 126

Answer 127

Answer 128

Answer 129

Answer 130

Answer 131

Answer 132

Answer 133

Answer 134

Answer 135

Answer 136

Answer 137

Answer 138

Answer 139

Answer 140

Answer 141

Answer 142

Answer 143

Answer 144

Answer 145

	23	9
14	9	5
4/11	1	6	4
11	3	8	

Answer 146

	22	12
9	6	3
3/17	1	7	9
11	2	9	

Answer 147

	12	11
4	1	3
10/22	9	5	8
7	1	6	

Answer 148

	22	9
16	9	7
16/19	9	8	2
12	7	5	

Answer 149

	22	16
16	9	7
11/23	8	6	9
10	3	7	

Answer 150

	19	12
17	9	8
12/11	5	2	4
15	7	8	

Answer 151

Answer 152

Answer 153

Answer 154

Answer 155

Answer 156

Answer 157

Answer 158

Answer 159

Answer 160

Answer 161

Answer 162

Answer 163

Answer 164

Answer 165

Answer 166

Answer 167

Answer 168

Answer 169

Answer 170

Answer 171

Answer 172

Answer 173

Answer 174

Answer 175

Answer 176

Answer 177

Answer 178

Answer 179

Answer 180

Answer 181

Answer 182

Answer 183

Answer 184

Answer 185

Answer 186

Answer 187
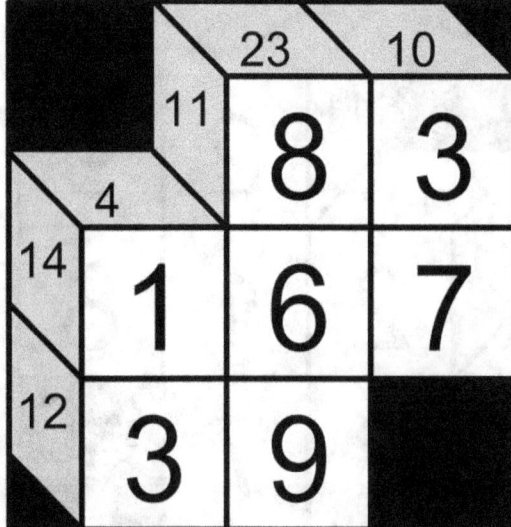

Answer 188

Answer 189

Answer 190

Answer 191

Answer 192

Answer 193

Answer 194
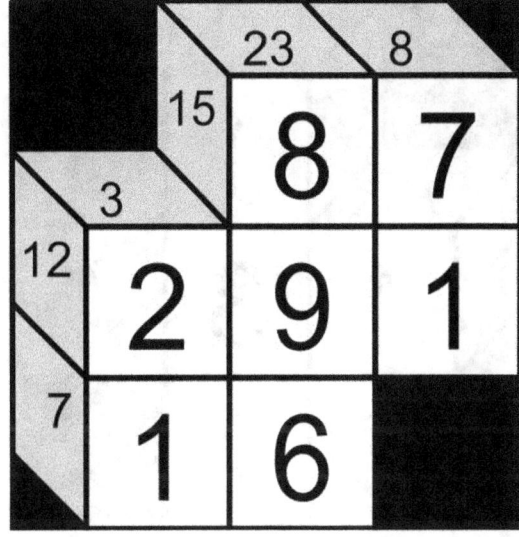

Answer 195

Answer 196

Answer 197

Answer 198

Answer 199

Answer 200

Answer 201

Answer 202

Answer 203

Answer 204

Answer 205

Answer 206

Answer 207

Answer 208

Answer 209

Answer 210

Answer 211

Answer 212

Answer 213

Answer 214

Answer 215

Answer 216

Answer 217

Answer 218

Answer 219

Answer 220

Answer 221

Answer 222

115

Answer 223

Answer 224

Answer 225

Answer 226

Answer 227
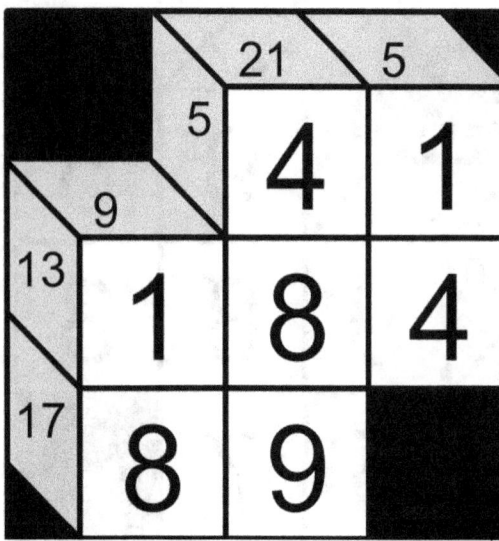

Answer 228

Answer 229
Answer 230
Answer 231
Answer 232
Answer 233
Answer 234

Answer 235

Answer 236

Answer 237

Answer 238

Answer 239

Answer 240

Answer 241

Answer 242

Answer 243

Answer 244

Answer 245

Answer 246

Answer 247

Answer 248

Answer 249

Answer 250

Answer 251

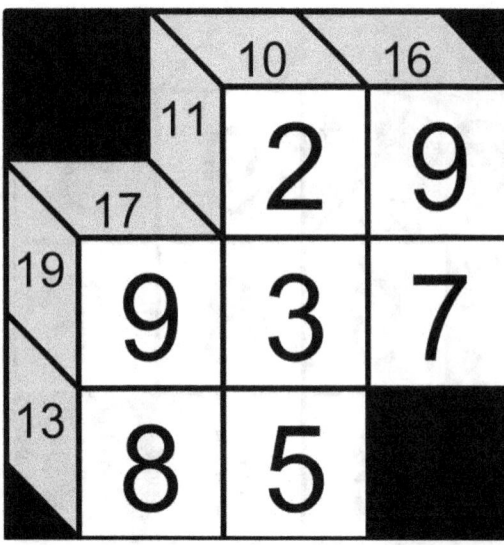

Answer 252

Answer 253

Answer 254

Answer 255

Answer 256

Answer 257

Answer 258

Answer 259

Answer 260

Answer 261

Answer 262

Answer 263

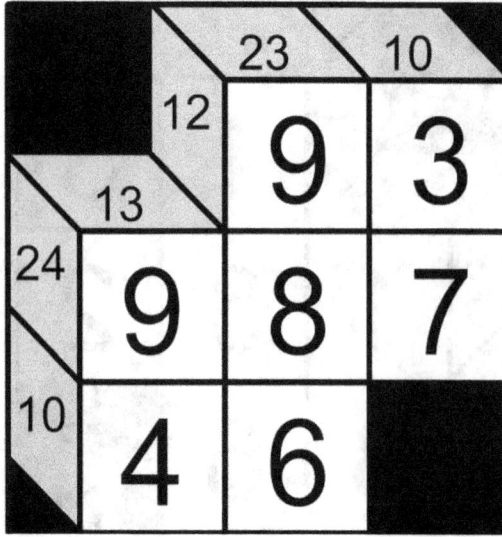

Answer 264

Answer 265

Answer 266

Answer 267

Answer 268

Answer 269

Answer 270

Answer 271

Answer 272

Answer 273

Answer 274

Answer 275

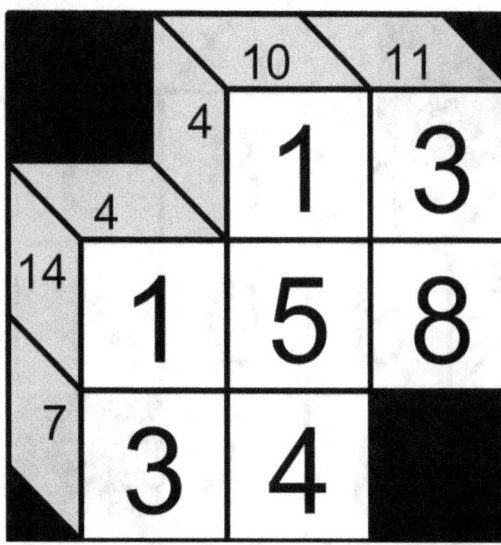

Answer 276

Answer 277

Answer 278

Answer 279

Answer 280

Answer 281

Answer 282

Answer 283

Answer 284

Answer 285

Answer 286

Answer 287

Answer 288

Answer 289

Answer 290

Answer 291

Answer 292

Answer 293

Answer 294

Answer 295

Answer 296

Answer 297

Answer 298

Answer 299

Answer 300